Thirlmere Water
– a Hundred Miles
– a Hundred Years

Norman Hoyle

Kenneth Sankey

Centwrite
Bury, Lancashire

1. The Roman Bridges with Great How and the Hellvellyn range.
Reproduced from a painting by permission of North West Water Ltd.

Text copyright © Norman Hoyle and Kenneth A Sankey, 1994

Published 1994 by Centwrite, 5 Hilltop Drive, Tottington, Bury, Lancs.,
as a fund-raising effort in support of WaterAid

ISBN 0 9523413 0 1

Typeset in Palacio, printed and bound in the United Kingdom by
Lamberts Printers, Station Road, Settle, North Yorkshire, BD24 9AA

Contents

Sponsors' Prefaces	5
Authors' Foreword	7
Now, and Then	9
— an explanatory walk	
— a look back in time	
Hesitation, Determination, Frustration	15
— a royal spanner in the works	
The Reservoir and the Dam	23
— plum stones and precipitation	
The Aqueduct	29
— a buried asset	
Water Quality	33
— keeping it clean	
Construction	37
— procedures, problems, provisions	
Some People	51
— they made things happen	
The Thirlmere Community	59
— always something happening	
The Land Around	69
— sheep, trees, visitors	
Aqueduct Development	75
— adding, linking, renovating	
A Century Passes	79
— millions and millions	
Appendix I	83
Appendix II	84
Centenary Greetings	86

Illustrations

1.	The Roman Bridges (reproduction of a painting)	2
2.	Thirlmere in the 1870's (reproduction of a painting)	20
3.	Engraved rock	21
4.	Centenary tree planting	21
5.	View of dam from downstream	22
6.	Thirlmere during filling	22
7.	Excavation at West end of dam	26
8.	'Plums' used in construction	27
9.	Facing of the dam	27
10.	The Swirls and Helvellyn Cove (reproduction of a painting)	28
11.	The dam as the trees begin to grow	28
12.	Tunnelling gang	32
13.	George Henry Hill	36
14.	Dunmail Raise	41
15.	The Reservoir overflow	41
16.	Straining well	42
17.	Construction village	42
18.	Aqueduct construction	43
19.	Pipelaying	43
20.	Short-span pipebridge	44
21.	Masonry viaduct	44
22.	Pipebridge—R. Lune	45
23.	Pipebridge—R. Irwell	45
24.	Aqueduct valvehouse	46
25.	Temporary siphon booster pump	46
26.	Overland pumping for reconditioning	47
27.	Aqueduct reconditioning	47
28.	Sir John J. Harwood	48
29.	Marble tablet	48
30.	John Frederic la Trobe Bateman	53
31.	Flood damage 1895	64
32.	Old Wythburn	65
33.	Class at Wythburn School	65
34.	Timber extraction	66
35.	Sheep dipping	67
36.	Ernest Brownrigg B.E.M.	67
37.	Strainer washing	68
38.	36" pipe burst Wardley Bridge	74
39.	Ceremony at Straining Well	80
40.	Thirlmere water comes to town	80
41.	Fountain in Albert Square	88

Front cover: Helvellyn across Thirlmere *Photo N. Hoyle*
Back cover: Aqueduct easement gate *Photo N. Hoyle*

Diagrams   Fig. 1. Map of Thirlmere 11
             2. Cross Section of Dam 14
             3. Cast Iron Pipe Joints 39
             4. Automatic Self-closing Valve 85

Most of the photographs of the construction stage were taken by E.H. Baldry who was the official photographer to Manchester Corporation Waterworks for the scheme.

Prefaces by the Sponsors of this Book

The People of the City of Manchester may feel justifiably proud of the celebrations to mark the centenary of the Thirlmere Water Supply and the present Officers and Members of the City Council are grateful for the opportunity to be associated with the publication of this book to commemorate the event.

The building of the dam and reservoir at Thirlmere to impound water and the construction of a one hundred mile long aqueduct to carry that wholesome water to Manchester are monumental achievements by any civil engineering standard. That they have withstood the passage of 100 years and are still serving the water requirements of this City's inhabitants, businesses and industries and similar needs elsewhere in the North West, is a fitting testimony to the memory of those who held a vision and laboured for its reality.

The authors, Norman Hoyle and Kenneth Sankey, are to be complimented on their efforts to recapture within the pages of this book all that is noteworthy yet leaving the interested reader eager to search for even more material on a fascinating subject. The City Council is privileged to commend it in the knowledge that proceeds will benefit the charity WaterAid whose primary objective recognises the great improvements to health and hygiene which wholesome water supplies can bring.

— Sinclair J McLeod,
City Engineer and Surveyor, Manchester City Council.

—o—

The Mott MacDonald Group, into which the firm G H Hill has been incorporated, are pleased to be associated with this commemoration of the Thirlmere centenary. The telegraphic address of that firm was for decades "Thirlmere", Manchester, an indication of the pride with which George Henry Hill and his immediate successors viewed their part in the creation of so great and important a project.

Mr Hill who, without the technology which has been developed during the past hundred years, succeeded in producing the Thirlmere Reservoir and Aqueduct, and numerous other waterworks facilities, would be delighted to know that his name is on the lips of present day engineers. Indeed, many lessons he learned, some the hard way, have been applied to become the bases of later works specifications. He would be proud also that the production of this book celebrating his greatest achievement will help to provide funds to aid the provision of water supplies for peoples less fortunate than ourselves—a most worthy cause.

— J A Turnbull,
Chairman, Mott MacDonald Group.

—o—

A century has passed since a delighted crowd watched the first water from Thirlmere gush from a fountain in Albert Square, Manchester. Much has changed during those hundred years but the Thirlmere reservoir remains a cornerstone of the North West's Water supply, and the aqueduct built by those Victorian engineers daily brings high quality water to Manchester and many other North West places.

In this book, Norman Hoyle and Kenneth Sankey tell how this noble work was conceived and carried out. They bring to life the memories of the men whose achievement it was: the far-sighted determination of the promoters, the sound judgement and leadership of the engineers, and the hard work and skills of the thousands of men who laboured to build the reservoir and the 100-mile aqueduct.

We at North West Water are proud to have the stewardship of this great asset. While raising the standards of water purity even higher, we have preserved the natural beauty of Thirlmere and have done much to ensure that it can be enjoyed by visitors from far and near. We are grateful to the authors, both of them distinguished civil engineers, for preparing this fascinating centenary history.

— Sir Desmond Pitcher,
Chairman, North West Water.

Authors' Foreword

In 1894 the dam at Thirlmere was completed, water impounded behind it, then first delivered into Manchester through the 96-mile long aqueduct. Having worked for the former Manchester Corporation Waterworks for over 20 years each, we feel that the centenary of the opening of this bold and enduring engineering achievement should be suitably recognised. So we have in this little book given an insight into the structural details of the dam and aqueduct and related the main happenings before, during and after the intensive first construction phase, while at the same time seeking to portray some of the human, economic and environmental factors.

We are gratified that the City of Manchester, North West Water and consulting engineers Mott MacDonald have seen our efforts to be worthy of their support and sponsorship, without which we would not have been able to proceed.

Compiling this record has heightened our admiration of the engineers, contractors and navvies who produced such a tremendous civic regional asset, not forgetting the foresight and determination of Manchester's Waterworks Committee of that time. We hope our readers will agree that we have done them justice.

We acknowledge the permission readily given us by North West Water to re-examine drawings and reports and to make use of the historical photographs in their care. We are also appreciative of the help given by Frank Oldroyd, Harry Whiteside and other erstwhile M C W W and N W W colleagues to fill some of the gaps in our knowledge and recollections of the Thirlmere reservoir and aqueduct, and to Ian McKenzie and his friends at Thirlmere for bringing to light facets of life there. Use of published source material is acknowledged on the appropriate pages.

The Lake District yields a clean and plentiful water supply to over three million people in North-West England, but others throughout Africa and Asia are less fortunate. So the knowledge that every copy of the book sold will help WaterAid to improve the quality of life for some community overseas is reward enough for us.

Norman Hoyle
Kenneth Sankey

April 1994.

"If Thirlmere had been made by the Almighty expressly to supply the densely populated district of Manchester with pure water, it could not have been more exquisitely designed for the purpose."

— Dr Fraser, Bishop of Manchester, 1878.

Now, and Then

An Explanatory Walk

Between 1890 and 1894 the City of Manchester built a small dam across a gorge on St John's Beck in Cumberland to create the very large Thirlmere Reservoir. It is easy to drive over the short dam without realising what lies beneath. For a close look, take a walk from the small car parking area at the road junction near the north west corner of the lake (O.S. map ref: NY 307 189).

Take the road which goes along the north side of the lake and from a bend the paved overflow channel can be seen down to the left. Where does it start from?—we shall soon see. At the rock cutting look for an engraving high up on the left which marked the opening of the new road to the public as part of the reservoir opening proceedings in October 1894.

Beyond the cutting look over the left hand wall to see the elegant curved line of the western section of the dam. Cross over and look back over the water for a glimpse of the overflow weir. The square valve tower is built up over what was a short term low-level overflow.

Next, a splendid marble tablet commemorates the laying of the first stone of this "embankment"; alongside it the Institution of Civil Engineers have added their own mark of appreciation. The valve house, from where the discharge through the tunnel beneath it into St John's Beck is controlled, stands behind the metal gates and is embellished with castellations and the Manchester coat of arms. Across the road can be seen the rock knoll which separates the two halves of the dam and through which the discharge tunnel was driven. Beyond the valve house is the taller, straight part of the dam whose foundation level is as far below the ground as the dam is high.

The only feature which appears to be lacking is some means of sending the water on its way to Manchester; but on the east shore of the reservoir, two and a half miles south of here, is the deep Straining Well at the entrance to the hundred-mile aqueduct to the city. It has another splendid "waterworks baronial" superstructure but the trees around it have almost screened it from view.

The wooded mini-mountain at the end of the dam is Great How and beyond it is the valley of How Beck which holds the scattered Thirlmere community.

Since April 1974, when water supply administration was re-organised, the Thirlmere reservoir, lands and aqueduct have been owned and managed firstly by the North West Water Authority and since September 1989 by North West Water Ltd.

In engineering terms Thirlmere is the perfect reservoir: substantial catchment area with high rainfall; flat bottomed valley allowing a large volume of water to be impounded by an unbelievably short dam; very clear and clean water; and an elevation sufficient to supply a city a hundred miles away without pumping. Thirlmere was in fact suggested to a Royal Commission in 1869 as part of a possible scheme to supply London from the Lake District but the members of the Commission recognised and reserved the potential of the Lake District to supply the large and increasing populations in the north of England.

A Look Back in Time

In 1875 the upper valley of St John's Beck in Cumberland must have been a very tranquil area. The flat bottomed valley held two lakes, Wythburn Water and Leathes Water (see Fig.1), joined by a shallow stream through water meadows and known collectively as Thirlmere. Helvellyn rising steeply on the east side and Armboth Fell on the west gave little room for habitations along the lake shores but a

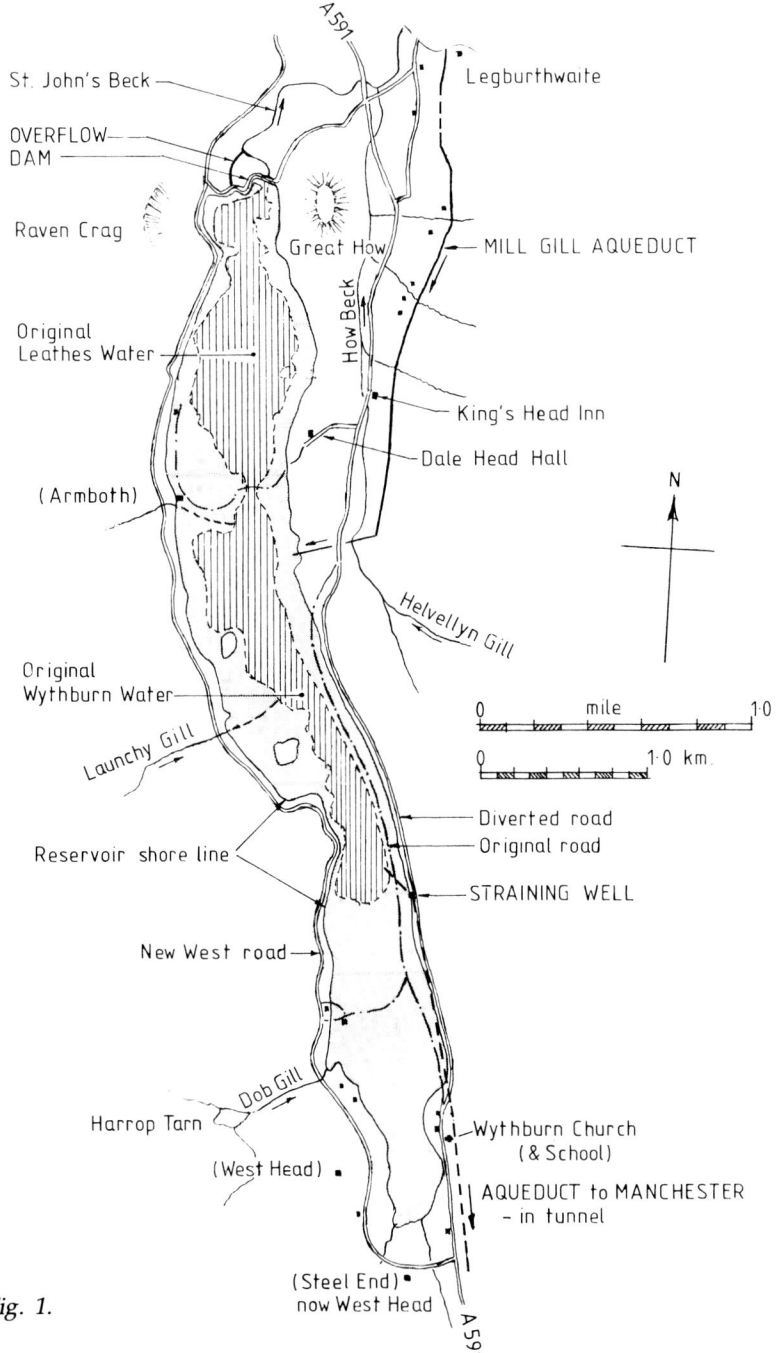

Fig. 1.

DEVELOPMENTS and ALTERATIONS at THIRLMERE

dozen or so farms and other dwellings were concentrated at the south end of Wythburn Water where the church was described in Wordsworth's poem "The Waggoner" as:
  "*Wytheburn's modest house of prayer,*
  As lowly as the lowliest dwelling."
Further north, Dale Head Hall, overlooking Leathes Water, was the home of Thomas Leathes Stanger Leathes, Lord of the Manor of Legburthwaite and owner of the lakes. The imposing mound of Great How beyond is featured in another Wordsworth poem: "Rural Architecture" telling of three boys who built such a large man-like pile of stones on its peak that it became known as "The Magog of Legburthwaite Vale".

Sheep farming was the mainstay of the valley, its conduct being regulated by ancient ritual administered by the courts of the lords of the manors of Wythburn and Legburthwaite. The valley was not however as isolated as some, the turnpike road from Ambleside to Keswick running along its eastern side with the toll bar cottage situated next to the vicarage at Wythburn. Though undoubtedly harsh in winter, it must have been an idyllic place on a summer's day; William Wordsworth and friends had a picnic here and, by popular belief, carved their initials on a stone by the lake side, there being no National Park byelaws in those days to deter them. When an attempt was made to remove the stone before it became submerged by the reservoir, it shattered but the fragments bearing the initials were assembled in a pillar alongside the diverted road, then "rescued" in 1984 and taken to Dove Cottage. A different opinion which has been put forward is that the initials were the work of a well known local prankster. Nevertheless, a plaque has been placed at the pillar from which the fragments were taken.

The first contact Manchester councillors made with this valley was an impromptu one bordering on disaster but worthy nowadays of a television comedy. A small party were acquainting themselves with Ullswater which their Engineer was advocating as the next source of water for the city. One summer's day, having made good progress with their viewing, they decided at Glenridding that instead of going back directly to Keswick they would walk the five or six miles over Helvellyn to Legburthwaite. After escapades of unaccustomed walking and horse riding, getting stuck in bogs, exhaustion, losing their way—and their tempers with each other, they eventually arrived at the Nag's Head at Wythburn late at night. Sir John Harwood's account of the journey is given in Appendix I.

The ordered lives of the land lords and their tenants must have been disturbed in 1876 by messages from afar that Manchester was contemplating a reservoir here, and their suspicions aroused by three gentlemen seeking details of properties and land holdings under guises of cattle dealer, builder and property agent, and tourist writer. This was part of Manchester's tactic of buying up as much as possible of the land and property by private purchase before promoting a Bill in Parliament, with the objective of reducing opposition to their plan.

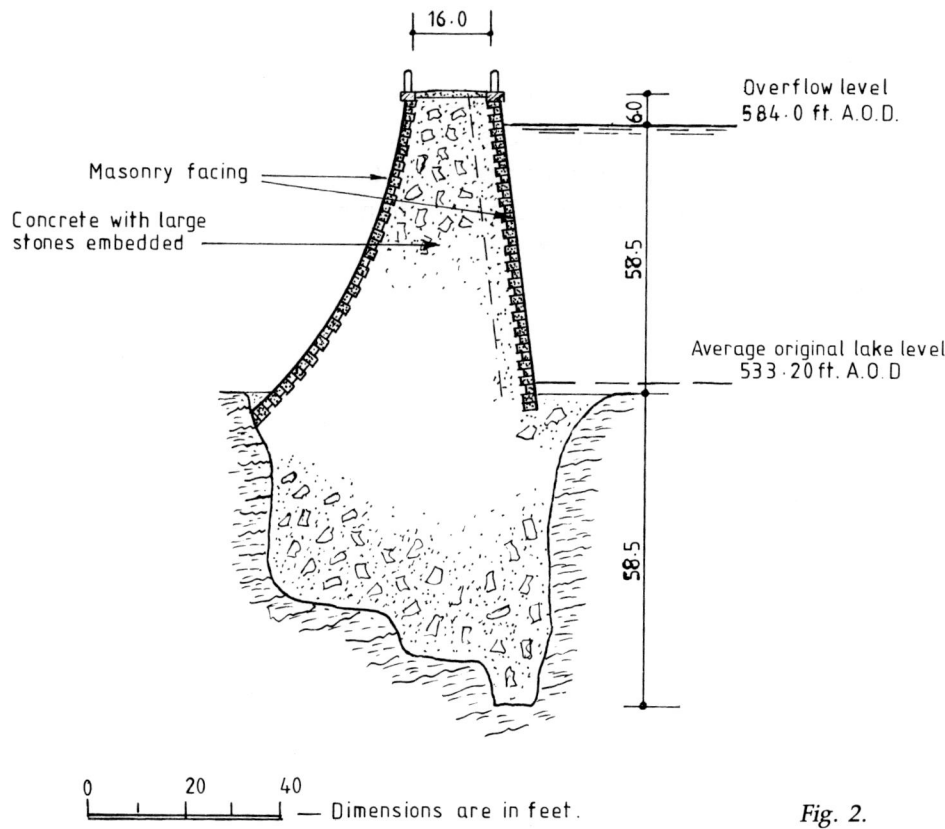

Fig. 2.

CROSS SECTION of DAM

Hesitation, Determination, Frustration

In October 1874 John Frederic Bateman, consulting engineer to the City of Manchester's Waterworks Committee, advised that the existing supply of water available from Longdendale, in a dry year, would be reliable to meet the rising demand for water in the city for only another seven years. In June 1875 he recommended to the Committee that they should seek powers from Parliament to obtain water from Ullswater, possibly augmented from Haweswater, either in partnership with Liverpool or independently. He had previously put the partnership idea to Liverpool but found them not very enthusiastic, an attitude readily reflected by Manchester. The Committee did however accept that something had to be done, especially after learning that fine cotton goods were being sent for bleaching and dyeing from Manchester to Glasgow where the clarity of the water was much superior. Bateman recommended that they should seek to develop a scheme sufficient for the next 20 or 30 years and yielding 40 or 50 million gallons per day; and, as there was *"nothing left in the hills from which you have hitherto gathered your supplies, nor in any district near at hand"*, they would have to look to the Lake District. Works of this magnitude and at such a distance would require at least seven years to develop after obtaining the necessary Act of Parliament; there was no time to lose.

The Waterworks Committee were not to be rushed. They certainly didn't want to join in any scheme with Liverpool. They did however resolve to see the Ullswater area for

Note: This chapter is drawn mainly from the book: "History and Description of the Thirlmere Water Scheme", by Sir John James Harwood, Chairman of Manchester Waterworks Committee, published by the City Council in 1895.

themselves and were *"much struck with the quality and important character of the residential property on its banks and the enormous cost that would have to be incurred should this property be submerged or injured in any way"*. For a time after the Committee had chosen to disregard Bateman's Ullswater proposal he became, at the age of 65, rather poorly and did not press the Ullswater-Haweswater suggestion. Then the Chairman, Alderman Grave, suggested that Bateman should look at Thirlmere; this he did and duly reported to the Committee very enthusiastically. *(Sixty years on, Haweswater was indeed developed as a water source for Manchester and was later augmented by pumping from Ullswater).*

The Committee set out to buy up the whole of the gathering grounds of the prospective reservoir, engaging people with local knowledge to make valuations but the councillors and aldermen themselves negotiated with the owners, so successfully that only one case, where an extortionate price had been asked, went to arbitration. Thomas Leathes Stanger Leathes, Lord of the Manor of Legburthwaite and owner of the twin lakes and a considerable extent of land, was steadfastly opposed to Manchester's efforts to buy the estate which had been in his family's ownership for over three hundred years. Conveniently for Manchester, he died while they were acquiring land and his son, who had emigrated to Australia, was quite prepared to dispose of the Leathes family interests in spite of continued opposition from other members of the family. On completion of the land purchases, the Chairman of the Waterworks Committee during his term of office became Lord of the Manors of Legburthwaite and Wythburn and it was incumbent on the Lord to keep a stallion, a bull and a boar for use by his tenants.

Leathes senior had threatened severe action against anyone acting for Manchester found on his land and Sir John

Harwood recalls how in the early days he and an elderly alderman crept around behind walls in the course of their viewing to avoid being seen. The city gentlemen had yet another chastening experience of venturing on to the fells. One beautiful afternoon—it must have been late spring— their experienced local guide suggested they walk up to the southern end of Helvellyn and down by moonlight. They had just the clothes they were wearing, no food or drinks, and when they got to the top they were struck by a blizzard. The guide declared that it would be dangerous to try to go down to Wythburn, so they made their way precariously down to Ullswater, arriving *"in a terrible plight"* at nine o'clock in the evening.

The Waterworks Committee were answerable to the City Council, but, right up to their demise in 1974, always enjoyed a good deal of autonomy; indeed places on this committee were eagerly sought and the chairmanship carried considerable prestige. Their recommendation in June 1877 to the City Council that a private Act of Parliament be applied for, to allow the development of a water supply from Thirlmere, was readily accepted, notwithstanding the likely enormous cost.

When Manchester gave public notice of their intention to construct the Thirlmere reservoir and aqueduct, this triggered off considerable agitation amongst many who were opposed to the idea of a reservoir in such a scenic area. Not content with stating a case for the preservation of the valley unaltered, they alleged that Manchester did not need the water, should go "somewhere else" for it, or do without; they branded the Corporation as speculators and profiteers and declared that Parliament was not competent to adjudicate the matter. One leading opponent was John Harward of The Hollens, Grasmere, who founded the Thirlmere Defence Association, but whose concern turned out to be the likely personal disturbance during the construction of the aqueduct through his land. He withdrew his opposition after the Waterworks Committee agreed to purchase his land and property, this being duly required by a clause in the 1879 Act. The Hollens is now the regional office of the National Trust.

A private Bill put before Parliament had to be examined by select committees of the House of Commons, and then the Lords, who considered opposition raised by objectors to safeguard their financial or property interests but seldom before, if ever, had a parliamentary committee been required to consider the impact of a scheme on landscape and outdoor enjoyment. The vociferous objections to Manchester's aspirations led to the Commons setting up a special enlarged "hybrid" committee to consider, inter alia, Manchester's water needs, the use of Lake District water and the water requirements of other towns between Thirlmere and Manchester. While representations were being heard, between 22nd February and 8th April 1878, Councillor Harwood and others held public meetings in Windermere, Ambleside and Grasmere to point out to "industrial people" of the locality the prospects of a local boom through trade and employment during the construction of the works, with the result that expressions of support for the proposed development were conveyed to the Parliamentary Committee in contrast to the sentiments of the intellectual gentry who opposed it.

After hearing evidence for and against Manchester on 17 days, the Select Committee reported back to the House of Commons that Manchester was fully justified in seeking additional supplies of water, that a period of 10 years was not a long time in which to develop a supply of the magnitude required and that it would be wrong of Manchester to appropriate any smaller water catchments earmarked by neighbouring towns or conveniently located to supply them. Thus there was ample justification for Manchester looking to the Lake District even though it was a hundred miles distant.

They accepted that Thirlmere was the most appropriate and practicable choice, rather than Ullswater or Haweswater, and made the following comment:

> "In the hills near Thirlmere the rainfall is about 100 inches annually. As the mountain barriers of this district form the chief natural condensers for the large amount of water which the currents of wind from the Atlantic bring to this island, it is a matter of public interest how far the water thus condensed can be used for the general public benefit. But the public at large has also an inheritance in the beautiful scenery of these mountains and lakes. The abstraction of water for public purposes should therefore be made subject to conditions which will confer its use without injuring the natural beauties of the Lakes and their surrounding mountains."

While the Bill was before Parliament, H R H the Prince of Wales (who became Edward VII) raised with the Royal Society of Arts the question of inequity of water supplies throughout the country in general, saying:

> "The smaller towns and villages are dependent on accidental sources of supply, and in many instances these are wholly inadequate for health and comfort. While the larger populations are striving, each independently and at enormous cost, to secure for themselves this article of prime necessity, the smaller localities must make the best shift they can, and in many instances are all but without any supply at all."

This royal observation, added to the representations brought by several Lancashire towns, prompted the Select Committee to introduce two additional clauses into the Bill, requiring Manchester to make "bulk supplies" of water available to towns and districts along or near the route of the aqueduct from Thirlmere to Manchester, on appropriate payment. Manchester did not contest this well-intentioned intervention which had unfortunate repercussions.

The passing on of the Bill to the House of Lords gave opponents a second chance to obstruct it and the Thirlmere Defence Association objected that the scope of the Bill had, through the addition of these clauses, been enlarged from its originally advertised purpose and the Bill was rejected for non-compliance with the relevant Standing Orders. Thus Manchester's plans were frustrated by a technicality over which they had no control. However, when an amended Bill was submitted to Parliament it had an easy passage and the Manchester Corporation Waterworks Act 1879 received the Royal Assent on 23rd May in that year.

2. Thirlmere in the 1970's. Leathes Water in the foreground, with Dale Head Hall below Helvellyn.
Reproduced from a painting by permission of N.W.W. Ltd.

3. Engraved rock commemorating the opening of the road over the dam
(see p.9). *Photo N. Hoyle*

4. Tree planting in 1979 to celebrate the Centenary of the Thirlmere Act
(see p.81). *Photo K. Sankey*

5. View of the dam from downstream. Note the valve tower and overflow discharge chamber. *Photo N.W.W. Ltd.*

6. Thirlmere looking North during filling. *Photo Abraham*

The Reservoir and the Dam

The water area of Thirlmere when full is 810 acres (3.25 square kilometres) and the volume of usable water stored is 8900 million gallons (40 million cubic metres). The dam at the north end which impounds this large amount of water is nowadays quite inconspicuous; in terms of reservoir volume created in relation to structural volume of the dam it must be one of the most effective and economical dams anywhere. It is in fact in two parts on either side of a rock mound at the central bend, its overall length being 857 feet (261 metres). The natural lake outlet was through the gorge on the east side of this mound, the river bed having been 68 feet (20.7 metres) below the present road level but the contractors had to excavate down through 58 feet below river bed level to establish a firm foundation. The foundation of the western part of the dam is at a higher level and its curving alignment was chosen for economy of construction. A cross section of the taller part of the dam at its original height is given in Fig.2; it was raised by 3.5 feet (1.1m) in 1918.

The heart of the dam consists of stone "plums" up to 3 tons in weight embedded irregularly in concrete. As the local slaty stone is difficult to dress to a smooth surface, the water side is faced with Pennine gritstone and the downstream side with Annan sandstone. Up to the time when the dam was built, the majority of reservoirs were formed by earth embankments, hence this solid wall of a dam was referred to in Corporation plans, contracts and papers as the "Embankment".

Excess water overflows over a two-part weir at the west end of the dam, through a tunnel beneath the road and down a wide masonry channel. A lower tunnel carried water away from an initial overflow, at the base of the rectangular tower, which gave a 20 feet usable reservoir depth, this being the depth to which the valley was first flooded. The overflow level was raised by 15 feet in 1904 and by a further 15 feet in 1916 to bring the water depth up to the authorised full depth of 50 feet. Shortly afterwards the overflow weir was raised through 3.5 feet to its present level and the dam wall raised correspondingly, this raising being duly sanctioned in a later Act of Parliament. Another tunnel, through the central rock mound, accommodates the pipes and valves through which the flow of compensation water into St John's Beck is discharged and larger discharges can be made, if required, to lower the reservoir level. The main valves controlling these discharges were originally operated by hydraulic pistons, the motive water coming from a tank high in the hillside; they are now electrically operated.

Water starts its two-day journey to Manchester through the Straining Well on the east shore of the reservoir, which is in fact at the southern end of the original Wythburn Water (Fig.1). Until 1980, when a more effective, automatic and compact rotary straining plant was built above Grasmere, water drawn from the lake was strained through a double ring of fine copper gauze panels which were raised individually for washing by a hydraulic crane fed from a never-failing spring in the side of Helvellyn. Also, as in the discharge tunnel at the dam, the original valves controlling the flow of water from the reservoir and into the aqueduct tunnel were hydraulically operated. After the panel strainers became redundant the opportunity was taken to streamline the flow from the lake into the tunnel by enclosing it in a pipe through the well.

The mountain peaks on the west side of Thirlmere rise to between 1930 and 2500 feet, those of Helvellyn and its neighbours on the east side to between 2750 and 3118 feet. Into this deep trough falls an average yearly rainfall of over 80 inches (2000 milimetres), but even this was not enough for

Manchester's engineers who cut and tunnelled a catchwater aqueduct two miles long to bring in water from Mill Gill and intervening becks to supplement the directly impounded catchment. Powers given in the 1879 Act to take water from Shoulthwaite Gill were never acted upon as the small additional amount of water would not have justified the cost of the tunnel required.

Since 1972 Thirlmere reservoir has been used in conjunction with pumped abstraction from Windermere to increase the yield. When there is a surplus of water in Thirlmere, taking into consideration the time of year and anticipated demand, water is drawn at a higher rate than the reservoir alone might be able to sustain over a long period. If Thirlmere becomes too depleted it is rested and water is instead pumped from Windermere, when lake conditions there allow, to Watchgate treatment plant and delivered into the Thirlmere Aqueduct north of Kendal. Thus the reservoir level now fluctuates more than in the past. In addition, since 1977 the rate of compensation water release into St John's Beck has been varied to improve the flow in the River Derwent.

7. Excavation at West end of Dam. *Photo E.H. Baldry*

8.   Note the "plums" in this construction photograph (see p.23).   *Photo Baldry*

9.   Facing progressing at the West end of the dam (see p.23).   *Photo Baldry*

10.   The Swirls and Helvellyn Cove across Wythburn Water.
Reproduced from a painting by permission of N.W.W. Ltd.

11.   The dam as the trees begin to grow.   *Photo N.W.W. Ltd.*

The Aqueduct

Water which goes all the way from Thirlmere to the Audenshaw reservoirs on the east side of Manchester travels 102 miles (163 km) at a speed of about two miles per hour. The Thirlmere Aqueduct is for a total of 36.75 miles like an underground or covered canal with a gentle fall of 20 inches per mile (1 in 3168). These conduit sections have a concrete floor and concrete or brickwork walls and arched roof, with inside dimensions 7 feet (2.1 m) wide and 7 feet high. The longest continuous length of conduit is between the rivers Lune and Wyre and is 4.35 miles long.

Where the contour being followed turns westward around a hill or mountain, the aqueduct goes straight through this in a tunnel. Where a valley disrupts the gentle fall the water is transferred across it through iron, steel or concrete pressure pipes but in a few places the conduit is carried over narrow valleys on masonry structures resembling railway viaducts. The tunnels are in general lined with concrete to the same internal dimensions as the conduits but some, where the rock is sound enough, have no lining other than a concrete floor. The water runs through a total of fourteen miles of tunnel, the longest being the first one from the reservoir, under Dunmail Raise, which is just short of three miles long.

The pressure pipe lengths of the aqueduct are known as siphons (though they are actually inverted siphons) and in general comprise four pipes in parallel as far south as Little Hulton, south-west of Bolton, but only three pipes across the shorter valleys in Cumbria. The three-pipe siphons comprise, in order of construction, pipes of 48-inch, 48-inch and 54-inch diameter, and the four-pipe siphons are 40, 44, 44, and 54 inches in diameter. The Act required that water must be carried across narrow valleys in the Lake District in underground pipes and not elevated on embankments or masonry structures, one exception being the crossing of Rydal Beck; unsurmountable difficulties did however lead to a bridge being built over Trout Beck. South of Kendal the tunnels are shorter and the siphons longer, the longest of the thirty siphons being 9.5 miles long across the Ribble valley and the greatest head of water, 427 feet or 185 pounds per square inch pressure, occurs alongside the River Lune.

Pipes are carried over the rivers Lune, Ribble and Irwell on handsome multi-span cast iron rib arch bridges and over lesser rivers on simpler bridges, but lie under the beds of smaller streams. The pipes are carried under railways and canals in subways except in the Manchester area where they were carried over the many railways formerly existing in steel box girder bridges, each with a drain to carry away any leakage water if a pipe joint should be disturbed by mining subsidence. Initially only one pipeline of each siphon was laid and further pipelines added as demand for water rose but the tunnels and conduits had at the outset to be made of sufficient size to carry the eventual fully developed capacity of the aqueduct. In addition, all valve wells, valve houses, subways and bridges were built large enough to accommodate the subsequent pipes.

At the inlet end of each of the thirty siphons is a "North Well" in which the admission of water into each pipeline is controlled. The wells were equipped with large and elaborate valves which would close automatically in the event of a burst pipe. Mr Hill's description is given in Appendix II. Likewise, non-return valves were fitted near the downstream ends of the pipelines to limit damage by backflow through a burst. In addition, massive intermediate manually operated sluice valves and ingeneous self-closing disc valves were provided on the longer siphons. All the early large valves were beautifully made to bespoke designs by

Glenfields of Kilmarnock and many of these, albeit with various modifications, are still functioning.

From Little Hulton the water was at first taken in twin pipelines across the Irwell Valley to a reservoir at Prestwich and these pipes were extended in 1926 to a larger new reservoir at Heaton Park. The Prestwich reservoir, after having become surplus to requirements, has been filled in recently and houses built on the site. Laying of the first of two pipelines around the south side of Manchester from Little Hulton to the large reservoirs at Audenshaw was completed in 1914 but the larger second pipeline was not added partway until 1952, then the remainder completed in 1968.

Mr Bateman's originally intended terminus for the aqueduct was to be in a reservoir to be built at Deane in Bolton, with pipes carrying the water forward into the Manchester system. This, to the proud City aldermen and councillors, would never do and he was persuaded to change the route on to a wider sweep around Bolton. However, in the strange way history has of catching up with events, in 1965 a balancing reservoir was completed at Lostock which is now part of Bolton.

12.

A typical tunnelling gang take a breather. Note the basic tools of their trade.

Photo Baldry

Water Quality

"If you will give us an unlimited supply of good and pure water, I will be responsible for the health of the city."

Dr Tatham, Medical Officer, Manchester, 1889.

The clarity of water in St John's Beck, coupled with the sparse population alongside it, was a significant factor in Manchester's choice of Thirlmere, a hundred miles from the city, as a source of water. Quality of water was in the 1870's judged largely on its colour and the clear water of the Lake District was in marked contrast to the peat-stained water from Manchester's Longdendale reservoirs. Little was known of water bacteriology but it was at least known that enteritic diseases could be spread by water contaminated from human and some animal causes.

Manchester had invested a massive amount of money in procuring a copious source of high quality water and it was logical that they should exercise strict control of all activities on the land draining into the reservoir. Hence their policy of acquiring ownership of the entire gathering grounds at the outset—a practice which the Government thirty years later was advising water supply committees, boards and companies to adopt. Perceptions of acceptable standards for public water supplies have since then narrowed progressively, even before the imposition of current EEC directive standards, and prevention of pollution and the degree of treatment necessary have been key factors in determining what activities may or may not take place on and around Thirlmere.

Most of the farmsteads were either inundated by the flooding of the valley, or loss of their valuable "inbye" land destroyed them as livelihoods; thus depopulation of most of the upper valley was a direct consequence of the reservoir. Some habitations in the Wythburn locality remained for several decades but all those around the church were eventually cleared in consequence of greater awareness nationally of the effects of pollution. Cattle were surprisingly allowed on the gathering grounds until 1918 after which only sheep farming was allowed.

Copper gauze "filters" in the Straining Well from the outset prevented particulate matter from being drawn into the aqueduct but after a few years it was discovered that the aqueduct was being coated with a fine deposit and two remedial measures were put in hand. Topsoil was removed from the lake margin and replaced by a covering of cobbles to reduce ground erosion by wave action, and in 1903 shallow settling pools were built alongside the aqueduct above Grasmere, on the east side of the road down from Dunmail Raise. These pools were doubled in area in 1922, to accommodate the increasing aqueduct flow, but proved totally ineffective for settling out the fine material; in fact they became a haven for gulls and water flowed back into the aqueduct in a more objectionable condition than when it entered the pools. After several decades of disuse they were filled in and the site landscaped.

To protect the conduit sections against damage by the external pressure exerted by ground water, pressure relief valves were incorporated in the floor, allowing the ground water to enter; in addition some spring and sub-surface water encountered was led into the conduit as a continuous flow. Over the years, with more intensive and mechanised farming adjacent to the aqueduct, the risk of contamination through these incidental inlets increased. To safeguard supplies arriving at Manchester, strainers were installed at Lostock in the 1920's and chlorination plant in 1944. Chlorine, a powerful bactericide, has been added to the water leaving the straining well since 1934 and lime added since 1944 to neutralise the slight acidity of the water. Some of the pre-1974 local water supply boards and departments declined to chlorinate bulk supplies taken from the aqueduct, arguing that it was incumbent on Manchester to deliver water which was already safe for drinking; a further problem was imposed by the numerous supplies given to individual rural premises as part of the bargaining over payment by Manchester for an easement for the aqueduct. However, following the unification of water supplies in 1974 these individual premises have been re-connected to local supply systems and chlorination has been applied to all bulk supplies.

The high quality of Thirlmere water does not demand the complex clarification and filtration process given to most upland-derived waters. As water treatment technology has advanced, improvements to equipment and control have been carried out and the micro-straining and chlorination now applied are deemed to provide sufficient lines of protection for the consumer.

13. George Henry Hill.
Photo by permission of Mott MacDonald Ltd.

Construction

The passing of the Act in 1879 was followed by a depression in trade which checked the rapid increase in water consumption in Manchester and a succession of wet summers left a surplus of water available from Longdendale. So the Waterworks Committee contented themselves during the next three years with completion of surveys, finalising land purchases and making wayleave arrangements for the aqueduct from Thirlmere but were in no hurry to embark upon the extensive and costly construction work. A severe drought in 1884 however depleted the Longdendale reservoirs to the extent that in November supplies had to be turned off for several hours each night, so in January 1885 they decided it was high time to make a start. Their first action was to dispense with the services of Mr Bateman, who had been their consulting engineer for some forty years but was now aged 75, and appoint as Engineer for the design and execution of the Thirlmere works Bateman's long serving assistant and partner for the previous five years, George Henry Hill.

Work was started first on the Aqueduct, and it was not until March 1890 that work at Thirlmere began. The contract for building the dam, together with diversion of the main road at Wythburn on to an alignment above reservoir top water level and construction of a new road along the west side of the valley was awarded to a Glasgow-based firm, Morrison & Mason, who were already driving the tunnel under Dunmail Raise; their tender amounted to £125,500 of which half was for the roads. Materials came by rail to Threlkeld and were then hauled the four miles to the site on wagons drawn by horse teams.

Notwithstanding mechanisation in the form of steam

cranes, locomotives, compressed air tools and concrete mixers, progress seems to have been laggardly at times, apparently due to labour disputes. Man power as provided by the British navvy was the contractor's principal resource but in September 1890 the Engineer complained that even the 570 men employed were not enough. He, for his part, seems to have under-estimated the depth to which weathered and fissured rock would have to be removed to reach a sound and watertight foundation. The additional foundation work took time and the dam which should have been finished by June 1892 was not actually completed until two years later. The Contractor claimed a total payment of over £184,000, in contrast to Hill's valuation of £162,366-6-0; a final settlement of £174,366-6-0 (£174 366.30) was reached and Morrison and Mason given a testimonial as to their efficient execution of the work.

The first contract for construction work on the Aqueduct, let in December 1885, was for the tunnel under Dunmail Raise and two other long tunnels at the northern end. Otherwise, construction work was apportioned into a number of comprehensive contracts awarded progressively from north to south, each comprising short tunnels, conduits, siphon pipelines, bridges and subways. The construction of the 96-mile aqueduct from Thirlmere to Prestwich was equivalent in its day to building a very long motorway. At peak activity more than 3000 men were employed by contractors on the sites in addition to those engaged in the manufacture of pipes, valves, bridge sections and basic building materials. The work was supervised for Manchester by over a hundred inspectors. For every yard of conduit, thirty tons of muck and rock had to be dug out, two thirds of this rammed back after the conduit had been built, and the rest carted away for tidy disposal.

Meticulous attention was paid to the design and construction of the thirty pipe siphons. Every one of the twenty five

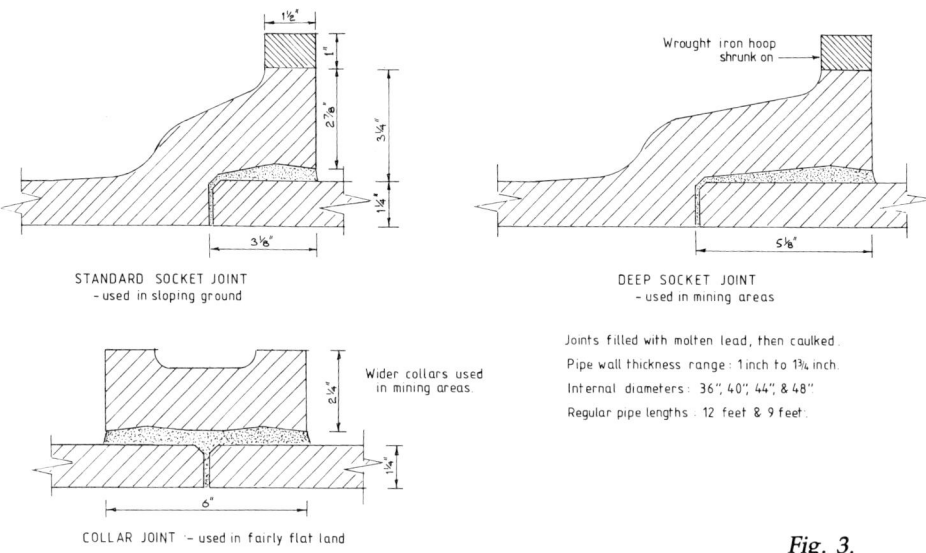

Fig. 3.

THIRLMERE AQUEDUCT
CAST IRON PIPE JOINTS

thousand or so pipes used in the first pipeline was numbered at the maker's works and the location in which it was laid recorded together with the name of the man who sealed its joints with hot poured lead. Pipes either had a socket at one end into which the machined spigot of the adjoining pipe was entered or were double-spigot joined by a slip-over collar. (See Fig.3). The insides of the sockets or collars were undercut to hold the lead in position against the internal water pressure. Curved castings were used for sharp bends but small or gradual changes of direction or gradient were achieved by use of double socket castings having each socket set at a slight angle. One of the principal suppliers of pipes for the first pipelines was the Staveley Coal and Iron Co., later to become Stanton and Staveley, then Stanton plc.

The contractor's tendered rates for driving the Dunmail Raise tunnel ranged between £9-0s-0d and £9-10-0 per yard driven, with an additional £4-0-0 to £4-10-0 for concreting the floor and walls. A typical rate for excavation for a conduit length, requiring between 15 and 18 cubic yards of digging per yard of trench, was £10-7-6. Excavation of the trench for the first line of Lupton siphon was priced at 7s-6d (37.5p) per lineal yard with 4s-6d (22.5p) per yard for pipe laying and 19s-0d for a pipe joint. All the contractors appear to have under-priced, all made claims for final payment well in excess of their tendered sums and had to wait up to four years for a settlement. One wrote to the Chairman offering to accept a considerable cut in his claim if he could be paid immediately by open cheque, remarking: *"As this contract has been such a losing one for us we cannot afford to lose even the bank charges any longer."* He was in due course engaged as one of the contractors for the second line.

During the laying of the first pipeline of the Grizedale siphon, near Garstang, a landslide occurred down the steep northern side of the valley, caused no doubt by the disturbance of unstable ground. The remedy adopted was to sink a shaft from the north well and drive a tunnel beneath the affected area, to contain this and future pipelines. One early defect in the aqueduct which had to be rectified arose in lengths of the conduit at Hutton Roof which in 1900 were found to be leaking. It was found that the surface of the concrete walls and floor was decomposing and in some places holes had been worn through; local limestone having a high lime content had been used in the concrete and had been dissolved by the soft Thirlmere water. The holes were plugged and the surfaces hacked back to firm concrete, then plastered with sand-cement mortar. This work had to be carried out during aqueduct shut-offs lasting thirty-six or forty-eight hours and such was the intensity of activity that up to 150 men were engaged.

To house the itinerant workmen and their women and children the Corporation and the Contractor erected 32 large huts in the vicinity of the works at Thirlmere. The Navvy Mission—an organisation dedicated to the moral and

14. Dunmail Raise, the first obstacle to be crossed on the journey to Manchester. *Photo N. Hoyle*

15. The Reservoir overflow. *Photo N. Hoyle*

16. The straining well house at Thirlmere (see p.24).   *Photo Baldry*

17. A construction village near Grasmere (see p.40).   *Photo Baldry*

18. Aqueduct conduit construction near Grasmere. *Photo Baldry*

19. Pipelaying on a steep gradient near Lancaster.
Photo by permission of Mrs Phyllis Harrison

20. Typical short-span pipebridge near Kendal. *Photo K. Sankey*

21. Typical masonry "viaduct". *Photo N. Hoyle*

22. Multispan pipebridge over R. Lune near Lancaster (see p.30).
Photo N. Hoyle

23. Pipebridge over the R. Irwell near Manchester. *Photo N. Hoyle*

24. Typical Aqueduct Valve house. *Photo K. Sankey*

25. Temporary Siphon booster pump with valve beam and canoe in background (see p.76 and Appendix 2). *Photo S. Bale Ltd.*

26. Overland pumping during Aqueduct reconditioning (see p.76).
Photo N.W.W. Ltd.

27. Reconditioning the conduit sections. *Photo N.W.W. Ltd.*

28. Sir John J. Harwood.
Reproduced from a painting in Manchester Town Hall by permission of Manchester City Council.

29. Marble tablet on dam (see p.9). *Photo N. Hoyle*

personal welfare of navvies and their dependants—made an early arrival and it was no doubt at their instigation that the Corporation and the Contractor contributed £100 each towards the cost of a temporary Reading and Recreation Room *"to be let out for various purposes tending to amuse and elevate the workpeople"*. The mission ran a simple hospital and provided a nurse. A school was built, serving also as a chapel on Sundays and as a soup kitchen when no work or wages were possible in winter. In the harsh weather of December 1891 the Corporation contributed £100 to a relief fund. The Waterworks Committee's minutes do not record the numbers of men killed and maimed but there was just a brief reference in the long and elaborate prayer at the opening ceremony to *"those who in sorrowful sickness and sore wounded gave up in this labour their lives"*.

As the aqueduct construction was more spread out than that at the dam, there was much less positive provision of accommodation and other facilities for workmen, even though more raw-boned navvying was needed. There was more scope for lawlessness and the Corporation were required to pay for the services of additional police in rural locations. The Navvy Mission was active as ever and the mission hut erected at Hutton Roof served as the village hall long after the navvies had gone. The story of the "Battle of Lupton" which took place a mile or so north of Hutton Roof has been passed down through families of navvies and goes:

"In September 1890, Lupton in Westmorland was the nearest village to two sections of the Thirlmere-Manchester pipe track. One was let to a Liverpool firm, the other to a Dublin company with an all-Irish workforce. Between three and four hundred men worked within drinking distance of the village. Most were single or temporarily womanless.

"They drank non-stop all over a weekend and through Monday. That afternoon an Irishman beat an

Englishman in a straight fight in the Nook Tavern. Small squabbles then broke out spontaneously and sporadically until they exploded into an open riot in which the English mounted a full-scale assault on The Plough (a mile away) where the Irish drank in a specially segregated taproom. The landlady tried to lock out the English, now armed with iron bars and staves, but when she failed she bolted the door into her main building, then bolted herself!

"It was now eight o'clock in the evening and dark. Outside it was cold and a breeze rustled the tall black hedges. Inside, before the English came, it was hot and fuggy, loud with Irish brogue and song. Then the English crashed in, and tobacco smoke and navvies swirled and rolled in the oil lamps' glimmer. Men roared like wounded bull calves. The fight spilled out into the yard and the road, except for three men lying in blood, spit and sawdust on the taproom floor. In a lane a hundred evenly divided men brawled and swayed. One man died after being smacked about the head with a spitoon."

This narrative is taken from the book: "Navvyman", by Dick Sullivan; Coracle Books, London, 1983, with acknowledgement.

Some People

Sir John Harwood (1832-1906)

Sir John Harwood was one of those self-made Victorians who took great delight in running and organising public affairs. As the Manchester Guardian put it, from the time he was 34 years of age his life was merged with that of Manchester.

John James Harwood was born of humble parents at Oswaldtwistle and spent his early working years as a print works labourer and a coach driver before being trained in plastering and decorating. He moved to Manchester to improve his education at night schools and at 29 was a partner in a painting firm, then sole proprietor at age 33 of "one of the largest concerns of its kind in England".

He was elected to the City Council for Cheetham ward in 1866, elevated to Alderman in 1881, thrice Mayor in 1884, 87 and 88 (Manchester did not have a Lord Mayor until 1893) and during the last of these mayoral years was knighted. He served on Parks, Cemeteries, Libraries, Gas and Tramways committees but his great love and effort was with the Waterworks Committee of which he was Chairman for twenty years. According to a feature on him by the Daily Dispatch in 1903 he dominated this Committee, yet members considered it an honour to serve with him; indeed, after he had duly declared open the Thirlmere works they subscribed to commission his portrait which still hangs in the Lord Mayor's Parlour. In contrast to this domineering manner, he displayed great patience and persuasion in acquiring land at Thirlmere.

He was one of the prime movers who in 1891 persuaded the City Council to invest heavily in the Ship Canal undertaking which was floundering and became one of the

Corporation's directors of that Company. It is said that when he resigned as a director in 1894 the value of M S C shares fell by a million pounds.

The Manchester Guardian in his obituary stated: *"He was a vigorous and incisive speaker, a little impatient of opposition to plans he had carefully thought out—circumstances which may account for his having more commonly inspired respect rather than affection."* He sounds to have been just the powerful man needed for one of the biggest municipal schemes so far. His pride in the Thirlmere supply was reflected in the descriptive and narrative book which he compiled.

John Frederic la Trobe Bateman (1810-1889)

John Frederic Bateman, who added his mother's maiden name la Trobe late in life, was educated in a Moravian school under the precept that *"pupils should learn the important principles of the mechanical arts, both that they may not be too ignorant of what goes on in the world around them and that any special inclination towards things of this kind may assert itself with greater ease later on"*.

After serving an apprenticeship with a surveyor and mining engineer he set up his own practice in Manchester at the age of 23, one of his first commissions being to examine the causes and effects of floods on the River Medlock. He became engrossed with rainfall and floods and subsequently dispelled before the British Association the commonly held belief that less rain fell on high ground than on lowland. In relation to rainfall and flood calculation he can truly be said to have been the father of engineering hydrology.

Bateman's big opportunity in works design came when (Sir) William Fairbairn, an eminent water power engineer,

Some of the information relating to J F Bateman and G H Hill is taken from G M Binnie's book: "Early Victorian Water Engineers" by permission of the publishers, Thomas Telford Ltd, London, 1981.

30. John Frederic la Trobe Bateman.
Photo by permission of Mott MacDonald Ltd.

asked him at the age of 24 to survey the civil works required for a scheme devised by a group of millowners ("Commissioners") in Northern Ireland who subsequently appointed him Engineer for the design and construction. After his design and construction of Hurst reservoir for the Glossop Water Supply Commissioners he was engaged by other water commissioners and by waterworks companies including those supplying Bolton, Blackburn and Chorley; his career was well established. An unsolicited testimonial from a top Government official led to Bateman being engaged by Manchester Corporation in 1845 to find appropriate sources of water to replace those of the inadequate and inefficient Manchester and Salford Water Company. This led to the establishment in 1847 of the Corporation's Waterworks Department and the construction between 1848 and 1877 of the chain of reservoirs in Longdendale with Bateman as Engineer.

During the construction of the Longdendale reservoirs Bateman was, to use a modern expression, still on a learning curve and the dams and other works were harmfully affected both by floods and obscure geological features but he was frank with the Waterworks Committee about the setbacks and continued to enjoy their confidence. In 1859 he moved to London and his practice, both there and in Manchester, became quite substantial. He and Thomas Hawksley became the pre-eminent giants of reservoir and waterworks engineering, each becoming President of the Institution of Civil Engineers; Bateman during his term of office claimed that he had made as many reservoirs as any man living— about seventy or eighty!

After he had turned sixty he seems to have become more autocratic. When in 1874 the Manchester committee were lukewarm to his realistic advice that they should start to look for a substantial additional source of water, he wrote to the Town Clerk remarking that he *"had plenty to do without the trouble and anxiety of getting up a great scheme for Manchester"*. There was however no-one else, other than the 67-year old Hawksley, with the credentials to impress Parliament in the promotion of a major water supply scheme. So Manchester continued to employ Bateman, with the successful outcome being the 1879 Act empowering them to go ahead with the Thirlmere works.

George Henry Hill (1827-1919)

George Henry Hill, a native of Stockport, joined Bateman as a pupil in 1843 and subsequently played an important role in the design and construction of the Longdendale reservoirs. The ability which he displayed during this time in dealing with extensive difficulties arising from landslips, floods and foundation conditions, led to him being chosen to direct the construction of a substantial part of Glasgow's Loch Katrine water supply works which Bateman had designed.

On completion of this work he was put in charge of Bateman's Manchester office after his employer had begun to spend most of his time in London. He was responsible for the completion of the Longdendale works, which included the design and construction of Vale House and Bottoms reservoirs and the rebuilding of the troublesome Woodhead dam. When the works in that valley were finally completed in 1877 he was accorded the rare distinction for a mere "Assistant Engineer" of having his name included on the Waterworks Committee's commemorative tablet.

Although Bateman was the promoting Engineer for the Thirlmere project it was Hill who carried out the survey work and feasibility studies. He also designed reservoirs for Stockport, Oldham and Ashton-under-Lyne but was not given partner status until after the Act authorising the Thirlmere work had been passed in 1879. It was a great distinction when in 1885 Manchester appointed him sole Engineer for their big project.

While the Thirlmere works were under construction the Manchester Ship Canal, a private venture, was also being dug, and when the City Council injected several million pounds into the project to save the Canal Company from insolvency one condition imposed, doubtless by Sir John, was that Mr Hill should act as joint Engineer for the completion of the canal.

At the luncheon following the opening ceremony at Thirlmere, Canon Rawnsley, a former defender of the Lakes against Manchester and railways and subsequently a founder member of the National Trust, in proposing a toast to the Waterworks Committee, expressed regret that the Queen could not be present. Had it been so, he declared, instead of rising as he did with a knight (Sir John Harwood) to the left of him, he would have risen between two knights, for no man better deserved the recognition that knighthood gave than *"he who so unaustentatiously had laboured on behalf, first of all of the Loch Katrine water supply, and then of the great and beneficial supply of water for Manchester"*. Knighthoods have indeed gone to many who have produced much less.

After completing the Thirlmere reservoir and aqueduct he took two of his sons into partnership, Harry Prescot Hill in particular emulating his father's reputation. The dams designed by George Henry himself and later by G.H.Hill and Sons are notable for their fine appearance; these include, in the western Pennines, Castleshaw, Yeoman Hey, Kinder, Fernilee, Dove Stone, Errwood and Jumbles. The cut-off trench for Kinder dam, near Hayfield, is at its greatest depth 215 feet below ground level; the record drawing carries a note at that point: *"Bottom passed by Mr.G.Hill, August 20th, 1908"*, he being a mere 81 years old at the time. Nevertheless in 1909 he declined the prestigious office of

President of the "Civils" in view of his increasing years and no doubt the distance of London from his home at Ernocroft Hall near Marple.

During the World War I years the Hills were engaged by Manchester to develop the Parliamentary plans for the huge Haweswater proposal but George Henry died a few months before the passing of the Act in 1919 which allowed that scheme to proceed. The Thirlmere works and his other reservoirs stand as fitting monuments to a great engineer, but on a personal basis there is a window in his memory, installed by his two daughters, in the cloisters of Chester Cathedral.

Bishop Fraser (1818-1885)

James Fraser DD was a native of Gloucestershire who, after various church appointments in the South of England, was called to be the second Bishop of Manchester in 1870. He immediately endeared himself both to civic leaders and the ordinary people of Manchester and Lancashire and was without doubt the first bishop to engage in "missions" to working men at their workplaces, drawing large audiences for dinner-break gatherings. He remarked in a personal letter: *"I do so hate those conventional sermons to fashionable congregations upon whom you know beforehand you are not likely to make the slightest impression. Give me a church full of Lancashire artizans ten thousand times rather, or a meeting of workpeople in a shed."* A man so acceptable to both employers and employed was often called upon to mediate in wage disputes, always seeking to persuade against both strikes and lockouts.

Not surprisingly, the "Citizen Bishop" supported the merits of the Thirlmere proposal and felt obliged to do his duty for Manchester in a place which he dreaded—the House of Lords. He attacked those who opposed Manchester from afar: *"Dainty and witty gentlemen leading a quiet life in London indulge in cynical carpings at our expense... and tell us that it is a thing not to be heard of that one or two million people should fetch a prime necessity of life from a Westmorland or Cumberland lake."*

His statue—erected by public subscription—still stands in Albert Square. Prince Albert and other notabilities face eastwards to the Town Hall but the Bishop looks northwards—to Thirlmere?

The Thirlmere Community

A number of farms and associated livelihoods disappeared with the flooding of the valley but the operation and maintenance of the waterworks created a new and larger local economy which became further enhanced by the forestry enterprise. The number employed, including forestry workers, rose to a peak of over sixty in the 1950's but with progressive modernisation and installation of communication systems this is now down to single figures.

There is no actual village named Thirlmere. The spread-out community between Thirlspot and Legburthwaite grew along the How Beck valley which does not drain into the reservoir and hence there was no constraint upon habitation. Some of the temporary buildings put up during the construction of the dam remained in use afterwards—in 1905 the Waterworks Committee donated five pounds to the funds of the mission room—but none now remain. New houses were however built for permanent employees, most of these now being privately owned. During Manchester's tenure few young men left the valley as they were able to obtain work with the Waterworks who were regarded as good employers.

In addition to the personnel at Thirlmere itself, a substantial number of men—at times more than 50—have been engaged on surveillance, maintenance and repair of the Aqueduct. With the Corporation having small depots along the line and houses to accommodate linesmen and inspectors, it could be said that the Thirlmere community stretched all the way to Manchester. As at Thirlmere, the number of aqueduct personnel is now much reduced from former levels.

Wythburn church and school remained active, outside

Manchester's ownership, but in 1895 the Vicar of Wythburn made an impassioned plea in the Manchester papers for the £16 needed for improvements to the school, in response to which the Corporation gave land for closets, then two years later contributed £2 to provide a treat for the children attending this school on the occasion of the Queen's diamond jubilee. In 1928 the redundant vicarage was purchased from the Church Commissioners and demolished "to ensure purity of water" and for the same reason ten years later the churchyard was closed for further burials. The Nag's Head Inn, which stood across the road from the church, was closed in 1930 and pulled down. By 1936, when it was closed, there were only six children attending the school adjacent to the church and it too was eventually demolished, its site now being occupied by part of the car park.

The parish of Wythburn at the southern end of Thirlmere is almost completely depopulated but the little church, established 350 years ago, remains and Sunday afternoon services are held monthly in summer. In the 1970's the Carlisle diocese wanted to declare the church redundant and have the building sold but such was the outcry from local residents, frequent visitors and M C W W workpeople that a fund to maintain the church soon raised enough money to pay for urgent repairs and the closure proposal was withdrawn. Its simple serenity was violated in July 1993 when thieves removed the roof slates.

The King's Head at Thirlspot has developed from a wayside inn, known locally as "Trespatt", into a tourist hotel and nearby Dale Head Hall, now privately owned, is also an hotel. This house, the former manorial seat, remained high and dry above the raised water level and its use was by favour of the Chairman of the Waterworks Committee who became Lord of the Manors of Wythburn and Legburthwaite. During the month of August however it was available to the Lord Mayor of Manchester while his apartment in the Town Hall was re-decorated and his servants on holiday. Local children were bidden to be on their best behaviour when the Lord Mayor or the Chairman was in residence and there was a reserved pew for these dignitaries in Wythburn church, the only pew with a cushion.

In 1921 the "home-made" electricty supply to the sawmill at Legburthwaite was extended to dwellings, making this the first isolated community in Cumberland to be lit by electricity, although outlying houses still had to make do with paraffin lamps. By 1948 the generator at the dam had became worn out and supplies were taken instead from the electricity board. One respect in which life here has recently become more peaceful has been the banning of heavy goods vehicles from this stretch of the Windermere to Keswick road following the completion of the M6 motorway through Cumbria and the improvement of road A66.

There are still a number of local residents who clearly recall life and work from the 1920's onwards. [See footnote]. A postman walked daily from Grasmere in all weathers, delivering mail as far as the post office at Fisher Place where others took over; his successor became mechanised with the issue of a red G P O bicycle until mail became re-routed via Keswick and delivered from there by a man with a motorbike and sidecar. A creature remembered with mixed affections was the Sentinel steam lorry; it was a sooty experience for workmen who at times had to ride perched on the roof of the cab. Bridge End was an early centre for social and other activities, the Stephenson family setting aside a building for use as a schoolroom and Sunday school; concerts and dances with a "reet good" supper were held in their barn. In response to pressure by parents, the local authority built a school at Legburthwaite which doubled in the evenings as the village hall and is now the Youth Hostel.

(Footnote; Their recollections and tales are being collected for separate publication).

The present village hall originally formed part of the temporary construction village at Haweswater; it was dismantled in 1941 after the completion of the dam there and re-erected in its present place.

During World War II the Home Guard protected the Straining Well and the dam, gun emplacements being built on Hause Point and Raven Crag. Employees needing to enter these areas had special passes bearing their photographs. The Home Guard were also on duty at some of the bridges along the aqueduct and there are unsubstantiated tales of bombing near-misses of the Ribble Bridge. The fire damage in the 1940 blitz on Manchester would have been much greater if the Thirlmere Aqueduct had been severed and the water supply direct into Trafford Park disrupted.

Local residents have learned to live with wet weather. Four years after the "opening" of the reservoir:

> "The largest flood experienced in the Thirlmere district for many years past occurred on the 2nd of November 1898. The water which flowed from the mountains above the east road on the Helvellyn side of the valley carried with it large quantities of debris into this road, amounting to some thousands of tons, which stopped traffic for several weeks. Some of the walls were knocked down and many culverts blocked up. The stream called Raise Beck which also brought down an unusual quantity of gravel and stones, did considerable damage to the public road between Dunmail Raise and Wythburn. Practically no damage was done to the west road, and none to the Works".

A flood occurred in 1931 when debris from a cloudburst blocked the Mill Gill aqueduct above Stanah Farm, causing water to overtop the bank and rush down the hillside. The report sent to Manchester reads:

> "Such was the rapidity of the rush of the water that there was no time to open the front door of the

Stanah Farm house and the water entering through the back door had no outlet until it reached the level of the windows which were broken and prevented further rising of the water."

The highest daily rainfall recorded is 178mm on 30th October 1977. By contrast, the lowest lake level reached in a drought was 43 feet-10 inches down from the overflow cill in September 1984. Of the many snowy winters, 1947 seems to have been by far the worst when blizzards isolated the Thirlmere area for several days and paths dug during the days filled with snow overnight.

31. Clearing flood damage at Brotto Farm 10th November 1895. *Photo N.WW. Ltd.*

32. Wythburn School, the Church and the Nags Head.
Photo N.W.W. Ltd.

33. One of the last classes at Wythburn School.
Photo by permission of Mrs Mollie Swainson

34. Extracting timber from the forest. *Photo by permission of Mr John Airey*

35. Sheep dipping time at West Head Farm.

36. Ernest Brownrigg B.E.M. Shepherd for many years at West Head Farm. *Photo N.W.W. Ltd.*

37. Members of the Reservoir Maintenance team washing a strainer.
Photo P.W. Allonby, Ambleside

The Land Around

Sheep

In the interests of water cleanliness only sheep farming has been permitted in the water catchment areas and even the extent of this has been curtailed by afforestation. Restriction on other livestock was not applied to corporation-owned land which did not drain into the reservoir. The standard of sheep farming in both tenanted and directly managed farms has always been high with commendable investment in farm buildings.

The native sheep are Herdwicks, hardy against the severe weather and nimble on the mountain sides but not the most profitable of breeds. Cross-breeding gave some concern in the Lake District that the pure Herdwick strain might be lost but the Corporation's flock produced a supply of true Herdwick tups. The West Head flock during Manchester's ownership varied between 2000 and 4000, shepherded for many years by Ernest Brownrigg who was awarded the B E M in 1967. West Head, which since 1984 has been tenanted, is the only farmstead actually within the water catchment area, although flocks of other farms graze the fells within the catchment.

Before 1970, many of the younger sheep were taken in winter to lowland pastures in Cheshire or on Walney island but the building of wintering sheds at the farms has reduced the need for these "holidays". The two strands of wire carried by metal posts atop the wall around the reservoir have attracted much criticism, sometimes being mis-reported as being barbed wire to keep people out; their true purpose is to prevent sheep jumping over the low wall when being driven along the road.

One early land improvement work carried out off the

catchment area was the draining of Shoulthwaite Moss—before wetlands came into vogue—and crops of potatoes were grown there during the first world war.

Forestry

In 1908 the Corporation embarked upon widespread planting of trees on the land around Thirlmere. Afforestation of a reservoir catchment area was at that time something of an innovation and it was not until the 1920's that the Government recommended this use of reservoir land. Apart from being a cash crop, the woodlands reduce soil erosion which would lead to discoloration of the water. The Thirlmere plantations cover 1900 acres (770 hectares), the main species being Norway and Sitka spruce, Douglas fir and larch.

The rapid spread of forest areas in other parts of Cumberland from 1920 onwards eventually generated public criticism of the dark and regimented plantations and some of this flew in the direction of Manchester Waterworks who had for more than forty years suffered little adverse reaction to their many trees. In recent years, as areas of mature conifers have been felled in accordance with programmes agreed with the Lake District Special Planning Board, plantations containing a proportion of broad-leaf trees have been established and forest margins "softened". Although the majority of trees in the plantations are conifers, there are noticeable numbers of mature deciduous trees around the lake margin and alongside the dam, Manchester having been under obligation by the 1879 Act to plant these.

Felling of mature trees at Thirlmere began in 1960 but in January 1974 nature decided to speed up the process and a great gale devastated the Douglas Firs between Low Banks and Armboth, over 2000 cubic metres (70,500 cubic feet) of timber being damaged. Where hand saws were once used for felling and the logs dragged out by horses, power saws and specialised logging machinery have taken over, but care has been taken to align haul roads so that they do not become eyesores. Before 1974 all forestry work was with few exceptions, done by Waterworks men; almost all felling is now carried out by contractors who often bid for the standing timber.

The sawmill at Legburthwaite has been operational for over 70 years. It was originally powered by a Pelton wheel driven by water piped from a small reservoir in Mill Gill but in 1921 electric power from a hydro-generator at the dam, driven by compensation water, took over. The sawmill's initial output was in the form of fencing posts and rails cut from forest thinnings and sold to local farmers but during World War II production turned to pit props for the Cumberland and Lancashire coalfields. A pressure creosoting boiler remained in operation until 1993 when age overtook the pressure vessel and other components, bringing about the closure of one of the last sources of creosoted products in the area. Farmers travelled miles to here to obtain the "genuine article", examples of which when dug out after forty years were found to be as good as new.

Specially grown Christmas trees have long been a Thirlmere speciality, the peak year's production having been 40,000. Several areas of land have been allocated to Christmas tree production, mostly in positions where the low trees afforded views across the lake. For over forty years a tree nursery provided a supply of saplings to be grown on as Christmas trees and for forest planting; the site is now occupied by the amenity car park at Legburthwaite. A specially selected larger tree is still sent to Manchester every year to stand in Albert Square.

The Thirlmere forestry undertaking has always enjoyed a high reputation, begun in 1920 with the award of a gold medal of the Royal Agricultural Society. The prestige was further enhanced between the wars by the Head Forester, Johnston Edwards who had come from the royal estate at

Balmoral; it was he who developed the tree nursery. Then in 1993 a "Centre of Excellence" award was made by the Forestry Authority, the regulatory arm of the Forestry Commission.

Access and Enjoyment

Manchester Corporation's stern enamelled notices which were displayed prominently around Thirlmere up to 1970, carrying threats of utmost action by the Town Clerk against trespassers, gave a broad impression of total exclusion from all the Waterworks estate. In fact Manchester had complied with the 1879 Act requirement that access hitherto enjoyed on the mountains and fells surrounding Thirlmere was not to be interfered with. Several footpaths to the summit of Helvellyn and those over to Watendlath remained freely open and people were free to wander on the fells. The notices also goaded a minority into clamouring for unrestrained availability of the lake for their own chosen form of recreation. Manchester replaced these with polite notices in 1970.

Steady improvements in water treatment allowed relaxation of the strict "keep out" attitude and in 1962 Manchester invited the Lake District Planning Board to say what degree of public access they would like to see. The response included suggestions for the provision of parking and picnic areas, lakeshore footpaths, camping and caravan sites and lakeshore access for boating. Manchester reponded swiftly by opening up the Swirls forest trail in 1963 and the Launchy Gill trail in 1965, each with a small car parking area.

North West Water Authority took over the dialogue in 1974, leading after much consultation to the publication in 1981 of a joint "Statement of Opportunities"; although the possibility of organised water sports was noted briefly, the main emphasis was on provision for walkers and "passive recreation". Interviews of visitors indicated that while everyone wanted access to and along the water's edge and more footpaths in general, no one wanted to swim, a minority wanted boating but no-one wanted to see or hear motor boats.

Many of the footpaths and car parks proposed in a further report in 1982 have now materialised, including a forest footpath from Dunmail Raise to Swirls with a fellside extension to Legburthwaite and one from Swirls along the lake shore and to the top of Great How. There are eight car parks of varying size and surface, with toilets at three of them; the smaller car parks are hidden among trees and the absence of approach signs makes it easy to miss their entrances! The need to avoid contamination of the reservoir and interference with farming and forestry has been well observed in creating these amenities. Nature conservation has also been given high attention, notably the red deer in and above the western side forests.

It is quite clear now that the eighty years or so of restricitve measures adopted by Manchester have resulted in the preservation of one of the few lakeside areas of unspoilt quiet in the Lake Distict.

38. 36" pipe burst Wardley Bridge, Swinton, 4th January 1956. *Photo N.WW. Ltd.*

Aqueduct Development

The tunnels, conduits, and the first siphon pipelines of the aqueduct from Thirlmere to Prestwich, totalling 96 miles, were brought into use on 13th October 1894. Further pipelines were added progressively to provide for increasing demand and some clever engineering measures were adopted to allow early use to be made of water from Haweswater and then to allow major maintenance of the aqueduct to be carried out after more than sixty years of continuous service. There has hardly been a year in the past hundred when some construction, maintenance or improvement work on the Thirlmere Aqueduct has not been under way. Some of the main activities are presented briefly below.

The second siphon pipelines were added between 1902 and 1904, followed by the third pipelines, completed in 1914. South of Little Hulton the third pipeline, instead of going to Prestwich, was taken all the way round the south side of Manchester to the Audenshaw reservoirs with delivery en route into Trafford Park. A long tunnel had to be driven to carry the third and fourth pipelines under the Manchester Ship Canal.

From 1922 to 1925 the fourth pipelines were laid as far as Little Hulton, the diameter in all of the siphons having been increased to 54 inches to obviate the need for a fifth pipeline. The pipes were mostly of steel with caulked lead joints but innovative reinforced concrete pipes were used on the Ribble siphon, these being made in a works set up at Longridge. In 1925 a booster pumping station was built at Lostock to drive water beyond Prestwich to a new, larger and slightly higher terminal reservoir at Heaton Park.

The Haweswater reservoir was built in the 1930's but, to defer for a few years the cost of a second long aqueduct to

Manchester, only the first nine miles of the Haweswater Aqueduct were driven and constructed, to a point where a short pipeline across the Sprint valley could feed water into the Thirlmere Aqueduct which had some spare capacity above the reliable yield of 36 million gallons per day from the Thirlmere reservoir. Construction of the Haweswater Aqueduct through to Manchester was delayed by the war and in 1947 pumps were installed at fifteen of the Thirlmere Aqueduct north wells to boost the flow of water through the siphons and increase the flow capacity of the conduits by reducing the build-up of water level at the siphon entrances. The pumps were two-speed axial flow, the higher speed drive being provided to give additional flow along a siphon if an adjoining pump or pipe should be out of service. They had electric motors which started and stopped automatically in response to the water level at the pipe inlet. The self-closing valves had to be modified to accommodate the pumps and their installation entailed the bringing of power to remote wells and construction of a transformer and switchgear house at each. The genius behind the pumping operation was George Eric Taylor, for many years Deputy Engineer and Manager of M C W W.

The first of four pipelines of the Haweswater Aqueduct was completed as far as Heaton Park in 1955, followed immediately by the laying of the second line. In 1957 the temporary transfer of water between the two aqueducts was reversed by means of a pumping station on the Sprint pipeline to reduce the demand on the Thirlmere Aqueduct. During the next three years major internal repair and reconditioning of the conduit sections south of Kendal was carried out, involving extensive overland pumping past the emptied sections of conduit. Subsequent reconditioning of the tunnels and conduits between Thirlmere and Kendal did however at times require complete stoppage of flow out of Thirlmere.

The completion of the second Haweswater pipeline also created a much needed opportunity to carry out scraping and lining of the Thirlmere Aqueduct pipes, the capacity of which had become reduced through encrustation. Over a ten-year period, 190 miles of siphon pipes were cleaned and given a cement mortar lining in-situ, using an American process and equipment as there was no British system then available for pipelines of 36-inch diameter and upwards. The length of pipeline to be lined, generally between 1000 and 1500 feet, was first prepared by forcing a cleaning machine through by water pressure. A thousand spring steel scraper blades bolted on to a central core removed the encrustation from the pipe wall and stiff wire brushes left a surface ready to accept the mortar lining. This was applied by another machine with a high speed centrifugal spray and spring loaded rotating trowel arms gave a smooth finish. The machine was driven by electric power and controlled by an operator lying in a forward cradle, the thickness of the lining applied being controlled by the speed of travel. The contractor for this work was Mowlem-Centriline but Manchester can claim credit for having promoted in Britain what has become a common water industry process.

In 1965 a covered service reservoir at Lostock was brought into use to give a balance between the steady flow down the aqueduct and the ever-varying rate of water use in the Manchester distribution system.

Laying of the fourth pipeline between Little Hulton and Stretford was begun in 1948, using 54-inch diameter steel pipes with welded joints and a bitumen lining. Soon after he had started work, the contractor withdrew and the Waterworks Department recruited engineers and a work force to carry out the work. Thus came into being the highly productive Waterworks direct labour works teams who subsequently laid most of the pipelines of the Haweswater Aqueduct and the pumping mains from Ullswater and Windermere, carried out the substantial conduit reconditioning and built seven large covered service reservoirs in the Manchester area. The engineers in charge of these works were John Keighley, Frank Oldroyd and the late Charles Hepburn. The second half of the fourth pipeline, to Audenshaw, was laid between 1964 and 1968 by the same direct labour organisation and this not only linked up with the Haweswater Aqueduct pipelines at Audenshaw to provide a most adaptable large diameter ring main around Manchester but also represented the final completion of the Thirlmere water supply system.

A Century Passes

Once the works at Thirlmere were complete, Manchester Corporation lost no time in arranging a grand opening. As the Prince of Wales couldn't come it was decided that the works would be declared open by none other than Alderman Sir John Harwood, Chairman of the Waterworks Committee. On 12th October 1894 members of the City Council and guests travelled by special train to Windermere, then by a cavalcade of coaches and wagonettes, led by the Lord Mayor, wearing his chain of office, through Ambleside and Grasmere to the Straining Well.

After a six-minute prayer, a short speech by the Lord Mayor and a long one by Sir John, he set in motion the hydraulic machinery of the valve to admit water into the aqueduct. What a moment! The culmination of four years work on one of the country's biggest civil engineering projects and an abundant supply of clean water now available to promote the health and industry of Manchester and its neighbours!

The following day, a Saturday, there was "a right grand do" in Albert Square. In front of local civic dignitaries, prominent citizens and a large crowd, Sir John was presented with an illuminated address and a gold key with which he proceeded to open a valve to start up a specially built fountain. The Police band played the National Anthem and the Town Hall bells pealed. Thirlmere water had come to town!

The arrival of Thirlmere water was not the only advance in Manchester in 1894. New Year's Day saw the first use of the Manchester Ship Canal which quickly established Manchester as a world trading port. In step with the increased water supply, the first deep interceptor sewers came into use to replace an inadequate and piecemeal system. Then

39. Official opening at Straining Well 12th October 1894.

Photo N.W.W. Ltd.

40. Thirlmere Water comes to town, 13th October 1894.

Photo R. Banks, M/cr

another Waterworks Committee development, the hydraulic power supply, started up, to drive the hoists of the warehouses of Cottonopolis and foster the development of manufacturing industry. The first hydraulic supply pumping station on Whitworth Street West had to be supplemented within fifteen years by others at Pott Street and Water Street. At peak use, in the 'twenties, 2730 machines in 597 premises were supplied through 30 miles of mains in the city streets at a pressure of 1000 pounds per square inch (69 bar). Bursts of the mains (3 inch to 6 inch diameter) were rare but spectacular! After 1930 electric power gradually took over, then with the decline of the British cotton industry the system was closed down in 1972.

One centenary was celebrated in May 1979, that of the Manchester Corporation Waterworks Act of 1879 receiving the Royal Assent. A simple service was held in Wythburn church where those taking part included Bobby Welsh who was a joiner at Thirlmere and Chairman of the local Parish Council, Gerry Fitzsimmons who was Lord Mayor of Manchester and a former Chairman of the Waterworks Committee, and the Bishop of Carlisle. Thanks were given: *"for the precious gift of water and for the work and skill which makes it readily available to men..."* Local children planted trees and it is a measure of the time span between the passing of the Act in 1879 and the turning on of the water in 1894 that some of those children now have children of their own.

In the hundred years since October 1894, well over a million million gallons of water or over five thousand million tons has flowed down the Thirlmere Aqueduct, and on peak days some two hundred and twenty thousand tons which is equivalent to 7200 tanker loads down the M6 motorway or one every twelve seconds passing Forton Services.

Age, pressure and corrosion have understandably had their effect upon the condition of parts of the Aqueduct. The insidious perishing of concrete, weakening of cast iron pipes and corrosive ground attack upon steel pipes which have taken place now dictate extensive remedial and replacement action. Far be it from present-day engineers, with another century's accumulated knowledge of behaviour of materials, to criticise the tremendous achievement of Mr Hill, his assistants, contractors and manufacturers. In the midddle of the twentieth century, Government loans to water undertakings for pipelines had to be repaid over a period of forty years which was reckoned to be their economic life, yet the original aqueduct pipes have already lasted a hundred years. Indeed the main problem areas lie in some of the later pipelines.

To overcome these pipeline and other defects, and deterioration in the conduit and tunnels sections, and so render this regional asset dependable for well into another century, North West Water are planning to invest some £60 million over a 20-year period. As Test cricketers have been known to say, a hundred is merely half way to double century!

END

Appendix I

Sir John Harwood's account of a journey over Helvellyn in 1875

As the mountains forming the west watershed of Ullswater are the boundary of the east watershed of Thirlmere, we determined to make full use of the day and return over Helvellyn, and by way of Legburthwaite. We could only do this by walking or going on horseback, and we were of the opinion that we could make a better inspection if we walked. There was however one difficulty. Councillor George Booth, one of the party, was advanced in years, and as he could not walk very well, he had frequently to rest, which was the more alarming on this occasion as we had a considerable distance to travel, and night was coming on. For this journey only one horse was hired, as we thought it would be sufficient to carry the luggage etc. We had perhaps been an hour on the way when Mr Booth required to rest. He was therefore placed on the horse, and not being accustomed to riding had to be partly held on, as well as could be managed, by the attendant.

After a while Alderman Grave, who for his age was a good mountain climber, began to be tired, and had to take a turn on the horse, and as the horse could not carry either of these men and the luggage as well, Mr Pape and I had to take charge of the luggage. We were going but slowly up the mountain when we heard a strange noise in the rear, and found that the horse, with Mr Grave on its back, had walked on a pocket of peat, and sunk above its knees in the bog.

It was a most amusing sight to see Mr Grave holding on to the horse, with his arms round its neck from fear of falling into the bog. We managed at last to get him off, but what to do, or how to proceed to get the horse out, we did not know. The driver did not try for some time to assist it, and it was evident he wanted us to leave him, so that he could return home. After a time the horse was got out, but we had not proceeded very far before we found that Mr Booth would either have to be put on its back again or be left behind. By a determined effort we arrived at the top of Helvellyn, when the question was, how and at what point should we descend? Mr Grave said he knew, or had been told, that there was a miner's track leading down to Thirlspot, so we tried to find it, and he went forward with Mr Berrey to descend at a point he thought led to the track. This adventure, however, proved to be a rather disastrous one, as both Mr Grave and Mr Berrey had a fall, and received considerable injury.

Mr Booth, having the horse, would insist on going along the top of the mountain and making the descent at Wythburn; but after getting over Helvellyn Low Man, ready to descend, the driver would not allow the horse to go further. We had therefore to gather up the luggage and do the best we could. We had a very troublesome journey with Mr Booth, but finally arrived at the Nag's Head Inn, Wythburn, after a hard struggle, about ten o'clock at night, where we waited some two or three hours, and then procured a carriage and drove to Keswick.

Sir J.J. Harwood: History and Description of the Thirlmere Water Scheme.

Appendix II

Mr G.H. Hill's description of the North Well Automatic Valves

A simplified illustration is given in Fig.4

In each of the siphon north wells there is a valve which, in the event of a pipe bursting, closes automatically, thereby cutting off the supply. The well is divided into two main compartments. One of these, the valve well, which receives the supply from the aqueduct, contains the valves, together with an overflow. The other compartment is sub-divided into smaller wells, one for each line of pipes. Each of these wells contains a float, by which the valve commanding that line of pipes is set in motion.

The water passes from the valve well to each float well through a cast iron tube set in the concrete floor. The end of the tube in the valve well is turned up so as to present a horizontal circular orifice 56 inches in diameter, which is faced with gunmetal and upon which the valve closes; the other end in the float well is turned up in the same manner and is provided with vertical guides in which the float rises and falls. The sides of the tube in this well are cast with seven circular orifices, each 15 inches in diameter, through which the water passes into the siphon pipe.

The valve consists of two concentric bell-shaped vessels, the outer of which is 56 inches in diameter and the inner 20 inches in diameter, the latter being used only in charging the pipes. The outer bell is of cast iron and the inner one is of Delta metal. The tops of both bells stand above the surface of the water, so that they form equilibrium valves. The float is hung from a lever 18 feet long, having the fulcrum at one end and the float at the other, the bell valve being suspended between the two and distant 4 feet 6 inches from the fulcrum. The fulcrum end of the lever is attached by a Delta metal pin to a cast iron box built into the roof of arched recesses in the side of the well, the float end being slung in a guide by chains which pass over pulleys and carry the counterbalance weights. The float, which is made of copper, is 52 inches in diameter and is connected with the lever by a rod. Passing horizontally through the centre of the large bell, about 6 inches above the seating, there is a rectangular tube open at the ends, in the middle and on the upper side of which there is a circular orifice with a gunmetal seating on which the smaller bell valve rests when closed. This small valve is connected by a rod through the top of the large bell with a canoe-shaped casting working on trunnions on top of the lever. In this canoe casting, which is $58\frac{1}{2}$ inches long, there is a cast iron ball 9 inches in diameter and weighing 80 pounds, the rolling of which tilts the canoe and so opens or closes the small valve. The trunnions are not placed at the centre of the canoe but nearer the valve end, giving a leverage of about 3 to 1. In the event of a burst occurring on the pipeline the level of the water in the float well falls, the float falling with it closes the large bell valve and the supply to the siphon is by this means cut off.

In order to recharge the pipe slowly and give the air time to escape, the small bell valve is opened by rolling back the ball to the end of the canoe nearest the float; the canoe is thereby tilted on its trunnions and raises the small valve. The water, which has always free access to the rectangular tube in the large bell valve, now escapes through the orifice thus opened and falls into the tube in the floor by which it passes to the float well and so to the siphon. When the pipeline is fully charged the water begins to rise in the float well and lifts the float, thereby lifting the larger valve. Meanwhile the raising of the lever by the float causes the ball in the canoe to roll to the end nearest the valve thus closing the small inner valve which is no longer required, the water being now admitted to the siphon through the large bell valve. The valves are now once again in position to cut off the water should another burst occur.

Abridged from the Proceedings of the Institution of Civil Engineers, 1895-96, by permision of the Institution.

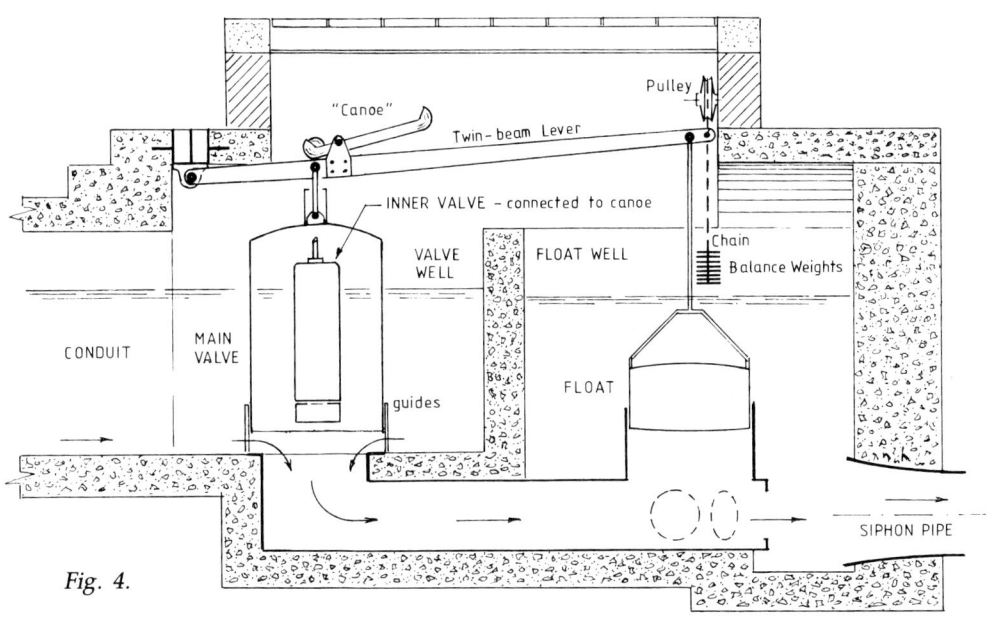

Fig. 4.

NORTH WELL AUTOMATIC SELF-CLOSING VALVE
Simplified Longitudinal Section

Congratulations Thirlmere
100 years and still running strong.

The following companies wish to be associated with the Thirlmere Centenary Effort and their generous donations have helped to increase the sum to be raised for WaterAid.

AMEC Civil Engineering Ltd. — Civil Engineering Contracting.

Aztec Environmental Control Ltd. — Manufacture and supply of on-line water quality monitoring equipment.

Babtie Group Ltd. — A multi-disciplinary consultancy offering commitment, experience, creativity, and effective management in all aspects of civil, mechanical and electrical engineering.

Binnie & Partners — A multidisciplinary consultancy encompassing civil, water and environmental engineering from a North West base at Chester.

Dale Head Hall Lakeside Hotel Thirlmere — For 100 years, residence Lord Mayor Manchester. Now idyllic Country House Hotel. Award winning cuisine served at the lakeside.

Dewsbury Civil Engineering Co. Ltd. — The Professional Specialist Utilities Contractor to the Water, Cable and Gas Industries.

J.F. Donelon & Co. Ltd. — Civil Engineering and Tunnelling Contractors.

Harbour & General Works Ltd. Morecambe — Building & Civil Engineering Contractors.

John Kennedy (Civil Engineering) — Mechanical & Civil Engineering Contractors.

Montgomery Watson — Specialist consultants in water, wastewater and the environment serving the world's environmental needs with over 2700 staff in 70 offices.

Norwest Holst Construction Ltd. — Civil Engineering & Utilities Contractor.

E. O'Donnell (Bradford) Ltd.	Management and construction of the rehabilitation of water distribution systems. Pipelaying in all materials and diameters.
Ozotech Ltd.	Specialists in the design and provision of complete ozone installations for the treatment of water, air, sewage and industrial effluent.
Rofe, Kennard & Lapworth	Consulting Engineers to the Water Industry since 1866. Designers of many dams in the North-West and responsible for safety inspections.
Sheffield & Company Ltd.	Forestry Contractors and Timber Merchants.
Stanton plc	Manufacture of Ductile Iron Water Pipeline Products and Municipal Castings.
Underpressure Engineering Co. Ltd.	Designers and Manufacturers of Pipeline fittings.
Wallace & Tiernan Limited	Manufacturers of high quality systems for disinfection by chlorination and other chemical dosing systems for drinking water.

The total sum donated by the above Companies was £935.

41. The Thirlmere Fountain in Albert Square, Manchester.
Photo A. Entwistle

—o—

All proceeds from the sale of this book are being donated to WaterAid to assist in their funding of low cost water and sanitation projects in Africa and Asia.

WaterAid,
1 Queen Anne's Gate,
London, SW1H 9BT
is a Registered Charity
No. 288701.